Energy
66

一起摇摆！
Swinging Together!

Gunter Pauli

冈特·鲍利 著

王菁菁 译

学林出版社
www.xuelinpress.com

丛书编委会

主　任：贾　峰

副主任：何家振　郑立明

委　员：牛玲娟　李原原　李曙东　吴建民　彭　勇
　　　　冯　缨　靳增江

丛书出版委员会

主　任：段学俭

副主任：匡志强　张　蓉

成　员：叶　刚　李晓梅　魏　来　徐雅清　田振军
　　　　蔡雩奇

特别感谢以下热心人士对译稿润色工作的支持：

姜竹青　韩　笑　杨　爽　周依奇　于　哲　阳平坚
李雪红　汪　楠　单　威　查振旺　李海红　姚爱静
朱　国　彭　江　于洪英　隋淑光　严　岷

目录

一起摇摆！	4
你知道吗？	22
想一想	26
自己动手！	27
学科知识	28
情感智慧	29
艺术	29
思维拓展	30
动手能力	30
故事灵感来自	31

Contents

Swinging Together!	4
Did you know?	22
Think about it	26
Do it yourself!	27
Academic Knowledge	28
Emotional Intelligence	29
The Arts	29
Systems: Making the Connections	30
Capacity to Implement	30
This fable is inspired by	31

一只橙子在观看肥皂的制造过程。他向一只正等着被混合到肥皂中的椰子抱怨。

"你知道吗？等榨出我的果汁后，他们就会蒸煮我的果皮。"橙子说。

"很疼吗？"椰子疑惑地喊道。

An orange is watching soap being made. He complains to a coconut, who is waiting to be added to the mix.

"Do you know that after they squeeze out my juice, the peel will be steamed?" says the orange.

"Does that hurt?" the coconut wonders out loud.

就会蒸煮果皮

The peel will be steamed

这还不是最糟糕的

The worst is yet to come

"不。幸好我们没有神经,所以我们感受不到疼痛。但这还不是最糟糕的。"
"还有什么能比压榨你的果汁、蒸煮你的果皮更糟糕的呢?"椰子沉思后说道。

"No. Thank goodness we don't have nerves, so we don't feel pain. But the worst is yet to come."
"What could be worse than having your juice pressed out and your peels vapourised?" muses the coconut.

"噢，最好的清洁剂隐藏在我的果皮中。需要不断地转啊转，至少转一个小时才能开始提取出来。"橙子叹道。

"你还含有清洁剂？"

"Well, the very best cleaning product is locked into my peel. And it will have to turn and turn for at least an hour before it's ready to do the job." The orange sighs.

"You also contain a cleaning product?"

你还含有清洁剂？

You also contain a cleaning product?

但却不是可持续的

It was not sustainable

"是呀。你看,当人类意识到棕榈油能进行生物降解时,就有越来越多人想用。于是数百万公顷热带雨林遭到了破坏。那时人们才发现,尽管这是可再生的,但却不是可持续的。"

"真是糟糕的消息!"椰子惊叫道。"我原本被种植在世界各地用于榨油,但是非洲棕榈树能产出更多的油,以至于我都不受欢迎了!"

"Yes. You see, when humans realised that palm oil was biodegradable, more and more people wanted it. And then millions of hectares of rainforest were destroyed. That's when people found out that although it is renewable, it was not sustainable."

"What bad news!" exclaims the coconut. "I was planted around the world for oil, but the African palm produces so much more of it that I became unpopular."

"噢，我是在巴西的橙汁生产商们意识到他们可以从我的果汁和果皮赚钱后，才被当作清洁剂广受欢迎。而这意味着可以赚更多钱！"

"这么说，你是用废物制造出的清洁剂……"椰子评论说。

"Well, I only became popular as a cleaner after Brazilian orange-juice makers realised they could make money from my juice and my peel. And that means making more money!"

"So you are a cleaning product made from waste …" observes the coconut.

用废物制造出的清洁剂

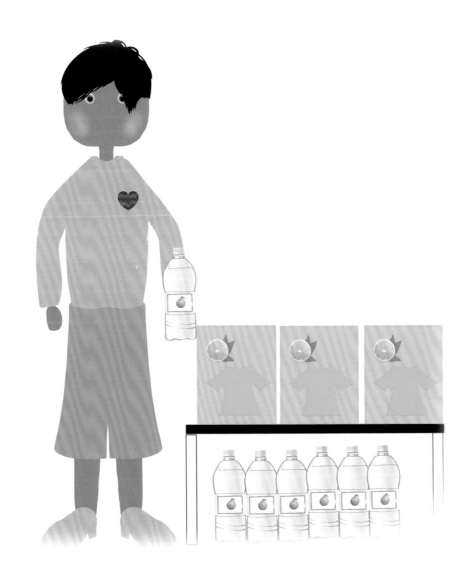

Cleaning product made from waste

我的整个果实都可以用

They use my whole fruit

"我不是废物。我的用途比人们原来认为的更多。我可以被用来制造果汁、果冻或者果酱。"

"噢,我也不只是产油这么简单。"椰子回答道,"他们用我制造一百多种产品。我的整个果实都可以用。"

"I'm not waste. I just have more to me than people think. I can be used to make juice, jelly or jam."

"Well, I am also more than just oil," responds the coconut. "They use me to make more than a hundred products. They use my whole fruit."

"咱们能提供这么多东西,真是太好了。要是人们能够充分利用我们,而不是只利用某个部分去赚钱就更好了。"

"不是有很多人在找工作吗?"椰子问。

"It's wonderful that you and I can offer so many things. If only people used us for all we are, instead of trying to make money off only one part."

"Don't they have a lot of people looking for jobs?" asks the coconut.

很多人在找工作

A lot of people looking for jobs

我们被转得晕头转向

We get dizzy from all that swirling around

"数百万人失业,数百万人想使用天然产品,而所有人都想节约能源。这就是我不明白为什么他们要扔掉这么多,还继续把我放在桶里转来转去的。"

"你是什么意思?"

"嗯,要想制造优质的清洁剂,他们需要将所有配料混合,而这个过程太漫长了。转来转去的,我们被转得晕头转向。"

"Millions of people are jobless, millions want natural products, and everyone wants to save energy. That is why I don't understand why they throw so much away, and keep on turning me around in vats."

"What do you mean?"

"Well, to make a good cleaning product, they need to mix all the ingredients and it takes so long. We get dizzy from all that swirling around."

"那么他们为什么不采用我听说过的一种新的强有力的摇摆?"

"你是指水流的自然摇摆,也就是漩涡吗?"

"是的。它很好玩!我们一起跳进去旋转,只要几分钟的狂野探戈,我们就都摇匀了。你应该尝试一下!"

……这仅仅是开始!……

"So why don't they use this new powerful swing I've heard about?"

"You mean the natural swing of water, called the vortex?"

"Yes. It's fun! We all get into a twist together, and within minutes of our wild tango we're all shook up. You should try it!"

...AND IT HAS ONLY JUST BEGUN!...

……这仅仅是开始！……

...AND IT HAS ONLY JUST BEGUN!...

Did You Know?

你知道吗？

Citrus peel contains the ingredients for cleaning products, perfumes, insecticides and solvents for 3D printing, but is usually thrown away.

橙皮含有制造清洁剂、香水、杀虫剂和3D打印溶剂的成分，但是通常都被扔掉了。

Palm oil is biodegradable and renewable but not sustainable since it is mainly farmed on land that was rainforest. This conversion from forest to agricultural land has destroyed the habitat of many primates.

棕榈油是可生物降解的，也是可再生的，但是却不可持续，因为它主要的种植地区过去原本是热带雨林。这种将森林转变成农田的行为已经破坏了许多灵长类动物的栖息地。

Oranges are the most cultivated fruit trees on Earth. Oranges keep fresh for up to 12 weeks after harvest and therefore were the preferred fruit on long voyages in days gone by.

橙子是地球上种植最多的水果。橙子摘下后可以保鲜长达12周，因此在过去它们是长途旅行所携带水果的首选。

Brazil produces more oranges than any other country in the world and exports 99%, most of it with ships the size of oil tankers.

巴西是世界上生产橙子最多的国家，其中99%用于出口，大部分用油轮一样大的船只运输。

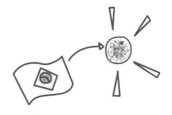

Coconuts are part of the daily diet of a billion people. It is a unique fruit that has more water than any fruit and provides oil for cooking, and making soap and cosmetics.

有十亿人将椰子作为日常饮食的一部分。它是一种奇特的水果，在水果中含水量最高，还能提供用于烹饪的食用油，以及用于制作肥皂和化妆品。

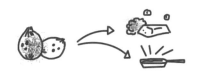

The name coconut was given by the Portuguese who thought that the shell of this nut looked like "coco", a witch in their folklore.

椰子（coconut）这个名称是由一位葡萄牙人起的，他认为这个坚果（nut）的外壳看起来很像他们当地传说中一位叫做Coco的女巫。

Soap is made from a vegetable oil, lye (sodium hydroxide or caustic soda, NaOH) and water. Homemade soap needs 4 weeks to cure, just like cheese.

肥皂是由植物油、碱液（氢氧化钠，分子式NaOH）和水制成的。家里自制的肥皂需要4周时间来硬化，就像制作奶酪一样。

To perfectly distribute all ingredients, oranges must be mixed and blended in a vat with an agitator, which takes time and energy. A vortex uses the force of moving water to mix.

要想让所有成分完全均匀地分布，橙子必须被放入一个大桶里用搅拌机混合调匀，这个过程消耗时间和能量。漩涡利用水流的力量来搅拌混合。

Think About It

想一想

Is it healthy to drink orange juice in the morning on an empty stomach?

早晨空腹喝橙汁是有益于健康的做法吗?

你会用一个女巫的名字或者原产国的名字给水果命名吗?

Would you name a fruit after a witch, or after its country of origin?

Why would one only press out juice from an orange when there is more money in the peel?

橙子的果皮含有更高的经济价值,为什么人们只用橙子榨果汁呢?

你想要种哪一种树:橙子树、油棕榈树还是椰子树?哪一种能够带来更多的好处?

Which tree would you prefer to plant: an orange, an oil palm or a coconut? Which one brings more benefits?

Do It Yourself
自己动手

This is a quick and easy one. Where in your house do you generate a vortex at least 7 times per day? Well, usually we go to the toilet 7 times a day, and each time you flush, you generate a vortex. This is an ingenious way to cleanse and transport waste: the process exploits the force of water flowing not more than 30 centimetres from the cistern. That is all that is needed to keep the toilet clear. Where else can you observe a vortex in your home? Go and check – you will realise that vortices are to be found all around. You might even have seen one through the window when a storm is raging outside.

这次是个快速简单的任务。在家里的什么地方你每天能制造至少7次漩涡？通常我们每天要去7次厕所，而你每冲一次水，你就制造了一次漩涡。这是清洁和运输排泄物的好方法：利用水从不到30厘米高的水箱流下时产生的动能，这就是保持厕所清洁所需的能量。在你家里还有什么地方能观察到漩涡？去找找看，你会发现漩涡无处不在。你也许曾经在暴风雨肆虐的天气里隔着窗户看到过。

TEACHER AND PARENT GUIDE

学科知识
Academic Knowledge

生物学	柑橘类水果极易通过不同种类的杂交创造出新品种；在人类体内，椰子油中的脂肪酸直接从胃部进入肝脏，然后转化成酮；可生物降解、可再生和可持续这三个概念是有区别的。
化 学	橙子是酸性的，其酸碱度（pH值）低至2.9，富含维生素C，不容易变质，可以预防坏血病；椰子富含中链饱和脂肪酸；橙皮里面的白色部分富含果胶。
物 理	肥皂使不能溶解的颗粒变得可溶解；肥皂具有亲水性，可以吸收湿气；钠皂是一种硬皂，用于洗衣；含钾肥皂是一种软皂，会产生更多肥皂泡，用于刮胡子；肥皂在水中产生电离反应形成碱性离子，使肥皂滑滑的；漩涡是自然界中水流运动的基本形式；科里奥利效应和地心引力决定了漩涡的旋转方向。
工程学	可以用蒸馏的方法从橙皮中提取天然化合物；漩涡是一种闭合的水动力系统，对水泵、螺旋桨、生物反应器和循环水管的设计有重大影响。
经济学	随机变量和概率分布对于一些商业领域的发展非常关键，如电信行业（多少个电话）、销售管理（每天多少名顾客）、机动车（多少辆轿车经过一个十字路口）以及生产率（需要多少酵母菌使酸奶发酵）。
伦理学	当有生产肥皂和食用油的替代来源时，我们怎么还能破坏猩猩的栖息地——热带雨林来种植棕榈树呢？
历 史	柑橘类水果最早种植于公元前4000年左右的东南亚，公元前2500年开始在中国栽种；橙子是由克里斯托弗·哥伦布引进拉丁美洲的；航海探险家们沿着航线一路种植橙子树；法国哲学家、科学家勒内·笛卡尔和英国物理学家开尔文勋爵都研究过漩涡。
地 理	多普勒雷达被用于寻找龙卷漩涡信号（TVS）；美国中部地区被称为龙卷风走廊。
数 学	随机变量和概率分布（如泊松分布）；平均值、变异系数、方差指数、平均差和标准差、中位数和高阶矩的概念；把水流运动看作非线性运动；水利工程过去是基于传统的伯努利方程和纳维-斯托克斯方程，但是现在是基于分形数学。
生活方式	个人卫生的文明始于用肥皂洗手来去除灰土和细菌；探戈舞（拉丁语意为"用针尖触碰"）兴起于阿根廷拉普拉塔河流域。
社会学	"碱性的"这个单词来自于阿拉伯语，意思是灰烬。
心理学	我们倾向于遵循逻辑路线，从一个论点到下一个，这种方式被称作线性思考；漩涡迫使我们用非线性方式去思考，从而得到我们从未期待或想象过的结果；跳舞可以让人们更放松，缓解低落情绪。
系统论	废物意味着损失了具有很多潜在用途的原材料；我们将废物转变成有用的产品和服务的方式需要得到大自然更多的启发，而且要更节能。

教师与家长指南

情感智慧
Emotional Intelligence

橙子

橙子对现状不满意,向椰子倾诉心事。橙子不是因为疼痛而抱怨,他是因为浪费了机会而泄气。橙子详细讲述了细节,并特别强调最困扰他的一点:混合搅拌需要很长时间。橙子具有现实主义精神,但是有自知之明,知道他的潜能只是需要得到开发。橙子意识到人类的无知,而且想知道他如何才能让人类认真思考对就业和天然产品的需求。最终,橙子成功地和椰子一起找到新的解决方法。

椰子

椰子对橙子表示同情,而且尽管她认可橙子的想法,她也没有抱怨,反而想办法来安慰橙子。她了解到他们有相同的作用,即清洁,但是她并没有把橙子看作竞争对手。椰子观察到他们两者之间的差异:橙子的清洁力来自于他榨汁后的剩余物。然而,椰子很有自知之明,声明她的好处不只是椰子油。椰子提出了一种更广阔的视角,即人类有就业的需求,她很贴心地让橙子解释她不清楚的问题。这种澄清激发了椰子的主动性,尽管面临艰难的处境,她仍选择以一种轻松的调子结束谈话,提议他们被混合在一起的时候一起跳舞。

艺术
The Arts

现在来发现铅笔的本领吧。画一个漩涡,但是直到你将漩涡的线条从起点一直画到纸边才抬起手。重复画这些线条,直到你画出漩涡的效果。漩涡经常是多个交织在一起,可以用一系列交织缠绕的线条来表现。交织的漩涡可以强化这种自然运动的力量。

TEACHER AND PARENT GUIDE

思维拓展
Systems: Making the Connections

当今的管理策略聚焦于具有核心竞争力的产业。这导致了产品专业化，每位生产商借此以独特工艺为基础来创建一个专门市场。任何无助于达成已严格明确的目标的事都会被看作是浪费时间和精力。这就是为什么生产橙汁的公司只对果汁感兴趣，并且会部署人力、科学和工程资源以更加低的边际成本来增加果汁产量。用橙皮的白色薄膜生产果胶、提取右旋柠檬烯和油以及生产溶解油脂的肥皂都是被忽视的机会，因为除了果汁其他东西都被看作无用的废物。这种策略会影响资源利用效率，而且意味着失去了很多工作机会。与此同时，现成的原材料被浪费了。如果有人能将以橙子加工为中心的各项活动集中起来，那么就可以创造更多价值，筹措资金提供额外的就业机会，并以更低成本提供更多产品。重新思考利用有效的可再生能源创造更多价值的策略，引出一个新问题：如果橙皮创造性地用于制作肥皂，为什么处理混合工艺要用一种过时的技术呢？液体、空气和天然气在自然界中都是通过漩涡来运输的，但是在工业中，这个过程却需要靠外力完成。创新胜过改变产品：工艺也需要改进。产品和工艺创新的结合展现了一条通向可持续发展的更快更长远的途径。

动手能力
Capacity to Implement

准备好制作肥皂了吗？找个哪怕把水溅出来一点（甚至很多）也没关系的地方。建议你戴上眼镜，以免有东西进入眼睛里。你需要一些材料：水、碱液和植物油。首先将碱液倒入水中，但是小心：千万不要反过来将水倒入碱液中！在小心地倒入碱液时，保持一段距离进行搅拌。因为这个过程会跑出些烟气，一定要确保你的脸和胳膊不会沾到。当水变清后，静置一会。加热植物油。现在要特别小心：这个时候所有东西都非常热，所以你需要戴保护手套。将这些东西混合在一起，搅拌至少5分钟。不要觉得辛苦，也许你能产生一个漩涡出来！用一块布盖住混合物来保温，然后放置一晚让它变成肥皂。如果第二天混合物变成固体，把它翻过来，用一个月让它进行干燥和硬化。四周后，把它切下来包好，这样就不会完全干透。你可以加点草药或者芳香油，创造出自己的特色。

教师与家长指南

故事灵感来自

科特·哈尔伯格
Curt Hallberg

科特·哈尔伯格年轻时不喜欢上高中，于是决定入伍加入瑞典海军。十几年里，他周游世界，在公海航行，在船上和海滨观察水流运动。从海军退役后，他在哈尔姆斯塔德技术高中学习开发工程。他对水流的运动感兴趣。然而，线性数学无法体现他以前观察到的水流运动的逻辑性和本质。他组建了一个团队，包括拉尔斯·约翰逊和莫滕·欧文森，一同研究水流的运行状况，特别是在瑞典马尔默市的实验室里研究水的净化。经过长达十年的研究、试验和技术发展，科特和他的团队创建了WATRECO公司，已经将两项利用漩涡原理设计的商业应用产品投放市场：一项是专为滑冰场设计的装置，能够用更少的能源生产更高质量的冰；另一项是利用漩涡的物理力将水软化，从而消除了对化学品的需求。

更多资讯

http://amrita.olabs.co.in/?sub=73&brch=3&sim=119&cnt=1

http://www.waterjournal.org/volume-4/pavuna#conclusions

http://www.vortex-world.org/rensnurr.htm

图书在版编目（CIP）数据

一起摇摆！：汉英对照／（比）鲍利著；王菁菁译．－－上海：学林出版社，2015.6
（冈特生态童书．第2辑）
ISBN 978-7-5486-0865-3

Ⅰ．①一… Ⅱ．①鲍… ②王… Ⅲ．①生态环境－环境保护－儿童读物－汉、英 Ⅳ．① X171.1-49

中国版本图书馆 CIP 数据核字（2015）第 086062 号

————————————————————————————

ⓒ 2015 Gunter Pauli
著作权合同登记号 图字 09-2015-446 号

冈特生态童书
一起摇摆！

作　　者——	冈特·鲍利
译　　者——	王菁菁
策　　划——	匡志强
责任编辑——	匡志强　蔡雩奇
装帧设计——	魏　来
出　　版——	上海世纪出版股份有限公司 学林出版社
	地　址：上海钦州南路81号　电话／传真：021-64515005
	网址：www.xuelinpress.com
发　　行——	上海世纪出版股份有限公司发行中心
	（上海福建中路193号　网址：www.ewen.co）
印　　刷——	上海图宇印刷有限公司
开　　本——	710×1020　1/16
印　　张——	2
字　　数——	5万
版　　次——	2015年6月第1版
	2015年6月第1次印刷
书　　号——	ISBN 978-7-5486-0865-3/G·314
定　　价——	10.00元

（如发生印刷、装订质量问题，读者可向工厂调换）